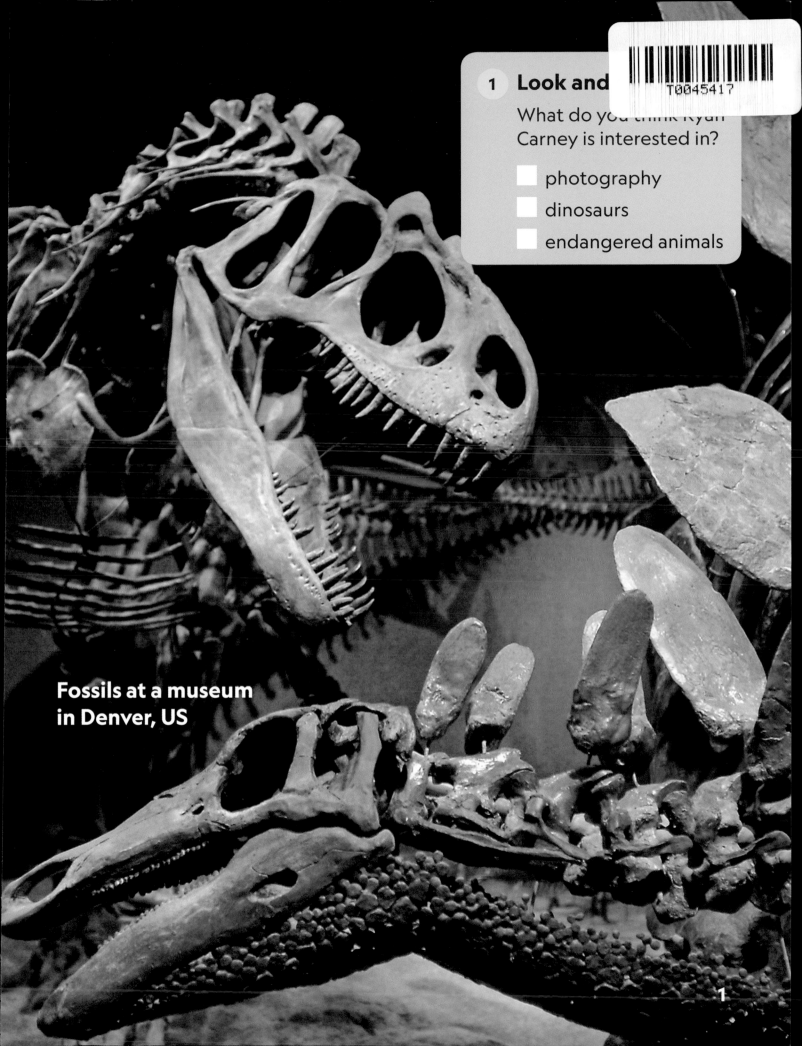

T0045417

Fossils at a museum
in Denver, US

VOCABULARY

1 **Learn new words.** Listen and repeat. TR:01

scan

fossil

computer program

animation

2 **Answer the questions.** Work with a partner.

Think of something you liked doing when you were little. Do you still like it now? Do you still do it now? Why or why not?

3 Read and circle.

When Ryan was young, he loved dinosaurs and birds. Today, he's a **paleontologist / fossil**. Now Ryan studies **computer programs / fossils** of dinosaurs.

Ryan wanted to **computer program / scan** a fossil of the *Archaeopteryx*. *Archaeopteryx* is a dinosaur that could fly. Ryan had to make a new scanner to do this.

Ryan likes computers, too. He uses **computer programs / paleontologists** to turn scans into **fossils / animations**.

4 Answer the questions. Work with a partner.

1 Have you ever seen a fossil? What was it of? Where did you see it?
2 Would you like to be a paleontologist? Why or why not?
3 Are you more interested in computer programming or animation?

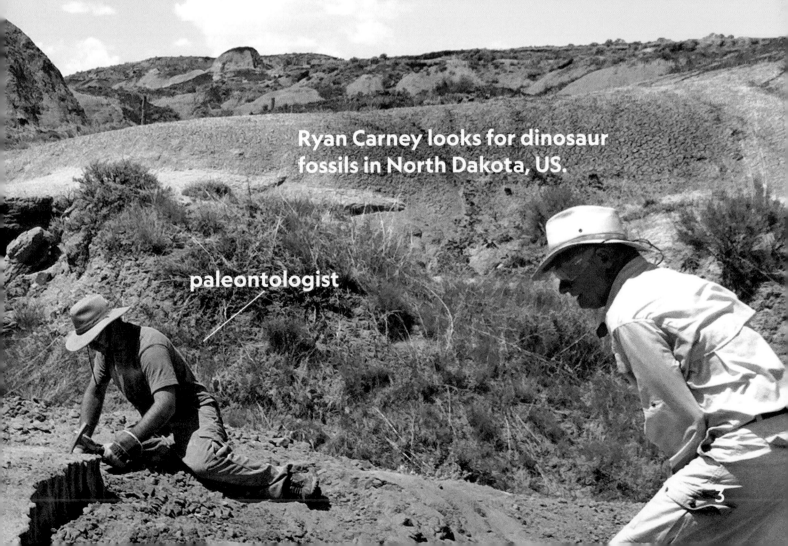

Ryan Carney looks for dinosaur fossils in North Dakota, US.

paleontologist

3

1 **Listen and say.** Then practice with a partner. TR:02

A: I like to draw dinosaurs.

B: I do, too. My class went to the dinosaur museum yesterday.

A: My class did, too. But we didn't see the *T. rex* bones.

B: We didn't, either. I liked the *Archaeopteryx* fossil best!

A: I did, too. My friend didn't. He doesn't like dinosaurs.

B: My friend Li doesn't, either! She thinks they're scary.

A: I think they're cool, but I don't want to be a paleontologist.

B: I don't, either. I want to be a teacher when I grow up.

A: I want to be an artist—and a photographer!

2 **Listen and check the correct answer.** TR:03

1 ☐ I did, too. ☐ I do, too.

2 ☐ The girls did, too. ☐ The girls didn't, either.

3 ☐ My father doesn't, either. ☐ My father does, too.

4 ☐ They don't, either. ☐ They do, too.

3 **Play a game.** Say what you like to do. Ask your friends what they like to do. Do you like to do the same things? Work with a group.

A: I like to swim. Annie?

B: I do, too. Duc?

C: I don't like to swim. Tram?

D: I don't, either. I like to dance.

4 **Write three things you both like and three things you both don't like.** Work with a partner.

A: I like video games. What about you?

B: I don't like video games. I like soccer. What about you?

A: I do, too.

like

1 _____

2 _____

3 _____

don't like

1 _____

2 _____

3 _____

5 **Play *Find someone who...*** Complete the sentences. Then ask your classmates. Try to find a classmate who did the same thing.

About me

1 I played _____ yesterday.

2 I didn't _____ yesterday.

3 I had _____ for dinner last night.

4 I didn't _____ last night.

5 I _____ to school this morning.

6 I didn't _____ this morning.

Classmate

_____

_____

_____

_____

_____

_____

A: I played soccer yesterday.

B: I didn't.

A: I played soccer yesterday.

C: I did, too.

Ryan Carney with a scanner and a fossil of *Archaeopteryx*

BEFORE YOU WATCH

1 **Talk about it.** Think of something you want to research. Work with a partner.

   1 What do you want to research?

   2 Why do you want to research it?

   3 What will you need to research it?

2 **Talk it over.** In this video, Ryan gives more information on his research on *Archaeopteryx*. What do you think he talks about? Work in small groups.

WHILE YOU WATCH

3 **Check the things that Ryan talks about.**

- [ ] *Archaeopteryx*
- [ ] birds' nests
- [ ] wings and feathers
- [ ] alligators
- [ ] a movie

- [ ] dinosaur habitats
- [ ] color cells
- [ ] a special microscope
- [ ] robotic cameras
- [ ] plants

AFTER YOU WATCH

4 **Read and circle *True* or *False*.**

1 *Archaeopteryx* had teeth, claws, wings, and feathers. **True / False**

2 *Deinonychus* had arms that could move and grab food. **True / False**

3 It took three years for Ryan to research how *Archaeopteryx* could have moved. **True / False**

4 Ryan didn't know what he wanted to do when he was young. **True / False**

5 **Ryan knows a lot about dinosaurs that fly.** What information about the dinosaurs did you like best? Why? Work with a partner.

> I liked the information about the feathers.

6 **Answer the questions.** Work with a partner.

1 Do you think it's easier to research a living animal or a fossil? Why?

2 What do you think are the difficult parts of a paleontologist's job?

3 Do you think researching dinosaurs is important? Why or why not?

READING

BEFORE YOU READ

1 **Learn new words.** Listen and repeat. TR:04

bone flexible protein steer

2 **Look at the photo.** What do you think the reading is about?
Work with a partner.

3 **While you read,** pay attention to the information about
flight feathers. TR:05

Amazing Feathers

Do you ever think about a bird's feathers? A small bird, like
a hummingbird, has about 1,000. A large bird, like a swan,
has 25,000. A bird's feathers can weigh more than its **bones**!

Feathers are made from **protein**. They're light, strong,
and **flexible**. A bird's muscles help the feathers move. Not
all feathers have the same job. Some feathers keep birds
warm. Some feathers give birds their colors.

Flight feathers help birds fly. They are on the wings and
the tail of a bird. Birds have large feathers on each wing.
They push the bird through the air. Smaller feathers closer
to the body help keep the bird in the air. Tail feathers help
a bird **steer** and make turns. They also help birds stop
and land.

Scientists now know that many dinosaurs had feathers
millions of years ago. The first birds, like *Archaeopteryx*,
lived 150 million years ago. In fact, birds today are living
dinosaurs! Think about that the next time you see a
chicken. Scientists think that there are now around 18,000
bird species, and each bird's feathers are special. Feathers
are amazing!

4 Answer the questions.

1 What are feathers made of?

_____

2 What moves a bird's feathers?

_____

3 Where are the flight feathers?

_____

4 Which feathers help a bird land?

_____

5 When did the first birds live?

_____

5 Talk it over. What did you learn about feathers? Work with a partner.

> I learned that tail feathers help birds steer.

6 Answer the questions. Work in a small group.

1 What is the most interesting bird you have ever seen? Was it flying? What did the feathers look like?

2 What bird would you like to see? Why would you like to see it? Where does it live?

"Remember to explore your curiosity... and never stop learning."

1 **Learn new words**. Listen and repeat. TR:06

award   contest   praying mantis   talented

2 **Listen and read.** TR:07

Ryan's Interests

You know Ryan Carney is a paleontologist, but he has many other interests, too.

Ryan started drawing when he was two years old. He loved to draw pictures, like this one, that combined nature and science. Ryan's art is amazing. He won many **awards** and **contests** when he was very young.

Now look at this **praying mantis** he drew when he was sixteen.

And this dinosaur!

Ryan is also an animator, a photographer, a sculptor, and a **talented** musician, too.

Ryan believes young people should do what they love and learn about the world around them. Just like him!

10

3 **Answer the questions.** Work with a partner.

What are your interests? Why are they important to you?

4 **Complete the sentences.** Work with a partner.

1 My family and I like to watch a popular singing _____ on TV.

2 I saw a _____ in our flower garden this morning.

3 My friend is a very _____ soccer player.

4 My class won an _____ for our recycling project.

5 **Check the things you think are important interests.**
Ask and answer. Work with a partner.

☐ playing sports

☐ learning computer coding

☐ watching TV

☐ cooking and baking

☐ _____

> I think playing sports is important. Exercise is good for us. How about you?

> I agree. It's also important to work together as a team.

6 **Ask and answer.** Work in small groups.

1 Look at Activity 5. Choose one interest that everyone thinks is important. Why do you think it is important?

2 What are popular interests for boys and girls where you live?

3 How can one of your interests help you when you grow up? Tell your group.

> I think cooking can help me. I want to be a chef. I can also cook healthy meals for my family.

Choose an activity.

A **Research a dinosaur or fossil.**

- Think of a dinosaur or fossil you're interested in.
- Find information on it.
- Write a report about it.
- Find or draw a picture of it.
- Present your research to the class.

B **Do an interview.** Work with a partner.

- Talk to someone who has an interesting job. Record the interview. One student asks questions. One student records.
- Ask if they wanted to do this job when they were young. Why?
- Find out interesting information about the job.
- Tell the class about this person.

C **Do a presentation on scientists.** Work with a group.

- Research and make a list of three different kinds of scientists. Write the name of a famous scientist for each one.
- Write an explanation of what they do.
- Present your information to the class.

EXTRA

What did you learn from your classmates? Write two things.

1 _____

2 _____

Play the game. Use a coin. Work with a partner.

Use the word to say one thing about Ryan and dinosaurs.

Use the word to say one thing about Ryan's other interests.

Move 1 space. Move 2 spaces.

START

paleontologist ► drawing ► *Archaeopteryx*

sculptor ◄ fossil ◄ awards

scan ► interests ► **FINISH**

PROJECT

1 **Make a vision board about your interests.**

- Think of three interests you have.
- Write them on a piece of poster paper.

- Find photos or draw pictures of your interests.

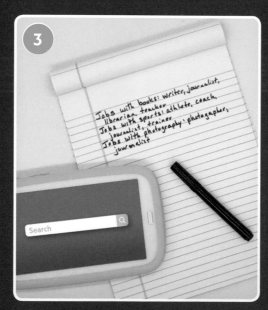

- Think of a job you could do with two or three of your interests. Do research to help find ideas.

- Find a photo or draw a picture of the job.
- Write about your interests and the job.

My Vision Board

I love to read books.

I love to play sports.

I love to take photos.

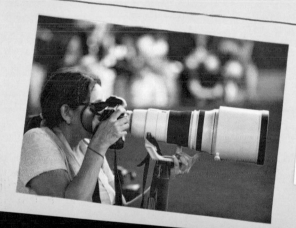

I want to be a Sports journalist. I can take photos and write stories about sports.

2. **Present your vision board to the class.**

3. **Say one thing about these subjects.** Work with a group.

- ☐ Ryan Carney
- ☐ *Archaeopteryx*
- ☐ studying dinosaurs with technology
- ☐ feathers and flight
- ☐ interests and curiosity

GLOSSARY

animation: a moving video made from a lot of images

Example: *The animation shows a dinosaur flying.*

award: a prize for doing something well

Example: *They got an award for cleaning up the park.*

bone: a piece of hard tissue in your body

Example: *There are 27 bones in your hand.*

computer program: instructions to a computer to tell it what to do

Example: *A video game is an example of a computer program.*

contest: an event to see who does something the best

Example: *She won a drawing contest.*

flexible: able to bend

Example: *Dancers are usually very flexible.*

fossil: something from an animal or plant that we can see in a rock

Example: *There are many fossils in museums.*

paleontologist: a scientist who studies ancient animal and plant life

Example: *A paleontologist often works outside looking for fossils.*

praying mantis: a large green insect

Example: *A praying mantis eats flies and other insects.*